Markets, Technologies & Opportunities:

Volume I: The Structure of The European High-Tech Economy

SERIES: Markets, Technologies & Opportunities:

Volume I: The Structure of The European High-Tech Economy

ISBN #:1-893211-99-1
Published: July 2017
©Paumanok Publications, Inc. 2017

Paumanok Publications, Inc.
244 Deerfield Road
Apex, NC 27523
www.paumanokgroup.com
Email: Info@Paumanokgroup.com

Markets, Technologies & Opportunities
Volume I: The Structure of The European High-Tech Economy

Table of Contents:

Markets, Technologies & Opportunities
Volume I: The Structure of The European High-Tech Economy

1.0 Introduction To The European High-Tech Economy:

The European high-tech economy is broken down in this book based upon regional analysis of serviceable geography, or basically a distance marker from airports or factories. As is the case in the USA, the market in Europe is fragmenting with demand being introspective and only remaining in place at this writing, and seemingly, because of protectionism; remains in the continent. The way to look at Europe is that everything seemingly that was not moved away from Europe has a supply chain that justifies its existence. This is exceedingly at the edge of the market where profitability is greatest (see visual model) below-

What Has Changed Since Our Last Review:

The European high-tech economy had been based upon a large footprint of wireless handset production in Germany in earlier reviews of this book; and this had shifted away from Germany but remained on the continent in former Soviet bloc countries, but this has tragically disappeared from the European high tech landscape and all component supply chains have largely followed suit. Therefore, the European high-tech economy now resembles a low volume, high value segment of the overall high tech economy that requires specific types of materials to continue to grow over the next five years.

2.0 Structure of The High-Tech Regional Markets In Europe:

The Central European High Tech Economy:

The Central European High-Tech Market Description:
The central European high-tech is 49% of regional European demand. This signature of material consumption supports a market rich in automotive and industrial end market customers and active factories.

Germany:

With a population of 81 Million people and 21% of the EU Economy, Germany is the largest consumer of materials supporting high-tech in Europe.

What is remarkable is that while Germany has increased its overall share of electronic component and materials consumption in the European continent since 2000, the overall value of component and materials consumption has shrunk due to the movement of key factories offshore to China, Korea and the Rest of Asia. Therefore the German market shifted dramatically to a dual fitting of automotive and industrial electronics accounting for the future direction of the German high-tech economy. The German market also shifted away from consuming mass produced low voltage components for the handset business model toward a high voltage component and material solution for the future automation business.

Bavaria, Baden-Württemberg and North Rhine-Westphalia Economies:

Three sub-regions of Germany are important for the reader to understand regarding a further breakdown of consumption in the country. The two southern states on the southern German border- Bavaria and Baden-Württemberg each accounting for about 21% of industrial output in the country, and the North Rhine-Westphalia region in the west of the country accounts for 20% of industrial output in the region.

The Complex German Market by End-Use Segment.

The German market for electronic components and materials can be easily broken down into three segments- Automotive & Transportation Electronics (57%), Industrial Electronics (15%), Consumer Electronics (10%) Power Supplies and Lighting (7%), Telecom Infrastructure (4%), Renewable Energy Systems (4%), Medical and Other (3%).

As we have stated this can largely be broken down into two segments- auto and industrial and all subsequent components will be in the low volume and high value realm of consumption going forward. Technical superiority is required for future goals in voltage, temperature and frequency to allow this model of future growth to move forward.

Three customers define European consumption of electronic components and materials- Siemens, Robert Bosch and Continental (VDO).

The Automotive "Under The Hood Markets" In Germany:

The German high-tech economy is largely influenced by demand for automotive electronic subassemblies which account for about 57% of the value of shipments for all components and materials consumed in country of which has much as 50% may be for "under the hood" engine control units requiring the most advanced technology for vibration, corrosion, and temperature withstand. This "Under-The-Hood" segment of the German automotive industry is very challenging, fast growing and profitable and largely centered at customers like Robert Bosch and Continental VDO in Germany.

Hungary:

Structure of The Rapidly Growing Hungarian Market for High-Tech Components and Materials

Thanks to Hungary's thriving high-tech sector, especially in the fields of "Information and Communications Technology", high-tech component and material consumption in Hungary has grown substantially. For decades, Hungary has specialized in vehicle manufacture, giving it an extensive pedigree in the auto parts and associated electronic components sectors. Currently, electronics manufacturing accounts for more than 20% of Hungary's total production output, while also providing about 112,000 jobs. Many world-class electronics companies have now established their regional headquarters, manufacturing facilities and R&D activities in Hungary.

This includes a number of global electronics manufacturers, such as Robert Bosch GmbH, *Electrolux, GE, Philips and Samsung*, as well as a number of the world's leading electronic manufacturing service (EMS) providers, notably *Flextronics* and *Foxconn*. In addition to Hungary's existing manufacturing facilities in the country, Samsung recently invested in a new TV manufacturing plant in Hungary. This has raised the company's total investment in the country to around 160 billion Hungarian forints (US$730 million).

The market in Hungary for electronic components and materials is substantial in value for FY 2017 and 9% of European components and materials consumption value and includes large quantities of back-end production markets in Eastern Europe for assembly of consumer audio and video imaging equipment and professional electronics.

In addition to factories previously mentioned that are operating in Hungary the large domestic Orion Industries is also in Hungary as well as is Videoton, a large EMS company catering to central European brands.

The Czech Republic:

The Czech Republic resembles the end-markets in Hungary- largely back-end assembly of consumer

AV and professional electronics and for automotive end markets.

The entire electronics industry accounts for more than 14% of Czech manufacturing output, which makes it the second largest sector in the economy. Over 17,000 companies employ more than 180,000 workers in the sector. Most of the sector's output is exported, mainly to markets within the European Union.

Key Czech Customers Buying Electronic Components and Materials:

Key manufacturers operating in the country include ABB (Power Automation technology), Olympus (Cameras), Delong Instruments (SEM) , FEI Company (Lab Equipment), Acer Computer (Computers), Tescan (SEM), Honeywell (Aerospace and Automation), Siemens (Factory Automation), Foxconn (EMS-Computers), Rockwell Automation, Eaton (Industrial automation and Auto Power Train), Panasonic (Consumer AV), Daikin AC (Big AC Plant), Denso (Auto Car AC Big Plant), Bang and Olufsen (Factory sold to Tymphany- makes high end audio) and Changhong (Consumer AV- Czech large home appliance factory).

Poland:

Consumption of electronic components and materials in Poland is largely known for TV set production. Poland also has unique expertise in the large-scale production of glass for cathode ray tube.

The author knows that certain US vendors of specialty glass source artisan glassworks in Poland, where the combination of skilled labor and low cost makes for a profitable global partner.

TV Set Production In Poland Drives Component and Materials Demand:

The industry is currently centered on the production of TVs, which accounted for an estimated 38% of electronic components and materials consumption in the country in 2017, and also, quite logically- computer monitors with a 24% share. According to official government figures in 2015 – about 20.1 million TVs and monitors were produced in the country, up from 19.6 million in the prior year, but down from the peak of 26.3 million in 2010. It is estimated that around 70% of production in 2015 were LCD TVs.

Key Polish Customers Buying Electronic Components and Materials in FY 2017:

The most important foreign investors in the electronics sector in Poland are: **Dell** (Łódź, Łódzkie Voivodship, production of desktops), **LG Display Poland** (Kobierzyce, Lower Silesian Voivodship, liquid crystal displays), **Jabil** (Kwidzyn, Pomeranian Voivodship, electronic components), **Sharp** (Łysomice, Kujawsko-Pomorskie Voivodship, production of LCD modules), **Funai** (Nowa Sól, Lubuskie Voivodship, TV sets), **LG Electronics** (Mazowieckie Voivodship, TV sets and other consumer equipment), telecom equipment manufacturers: **Alcatel-Lucent**, as well as **Kimball Electronics Poland** (Tarnowo Podgórne, Wielkopolskie Voivodship, electronic components for telecommunications and the automotive industry), **Flextronics International Poland** (Tczew, Pomeranian Voivodship, telecommunications components and products).

Slovenia:

1% of European component and material consuptionis in Slovenia and goes to a basket category of manufacturers who are catering to the production of small electrical motors for automotive, industrial and large home appliance products. There are large home grown electronic companies in Slovenia consuming components and materials such as **Gorenje, Kolektor, Iskra, Hidria, and Elektroncek.**

Slovenian Focus On Industrial Electronics:

Electrical and electronic equipment for the automotive industry, commutators for electric motors, electric motors, electronic components for household appliances, vehicle lighting, thermal management, explosion-protected electrical devices and wireless designs are among the products developed and manufactured in Slovenia largely for consumption by brand name high tech companies in Western Europe. Slovenia is an excellent choice for a private label partner in the electronic component industry because of its combination of skilled workers and low costs to produce.

Key Slovenia Customers Buying Components and
Materials in FY 2017:

In Slovenia we find the following fragmented
high tech component and materials companies-
Bartec (Explosion protected electrical devices),
BSH Finance & Holding (Domestic appliances),
Domel (Motors), **EBM Papst** (Small electrical
motors and fans), **Kona International** (Consumer
electronics), **E.G.O. Elektro-Geräte** (Electronic
domestic appliances), Gorenj **Panasonic
corporation** (Domestic appliances), **Hella KGaA
Hueck & Co.** (Vehicle lighting electronics), **Hdria**
(Motor vehicle electrical and electronic equipment),
Hdria (Electric motors, generators and
transformers), Bosch Rexworth (Electrical Motors
and Generators), **Elektroncek
Group** (Electromechanical gaming machines),
ISKRA* (Electrotechnical products for energy
sector and logistics); **Raycap** (Surge voltage
protection systems) ISKRA, ISKRA **El Sewedy
Cables** (Devices and systems for electric energy
measuring, registration and billing), ITW
METALFLEX **SG Invest Holding** (Components
for household appliances).

The Northern European High Tech Economy:

The Northern European Capacitor Market
Description: FY 2017:

The Northern European high-tech economy includes Holland, United Kingdom, Switzerland, Ireland and Sweden. Here the signature of consumption differs as the Central region because it prefers components and materials that supports a market rich in consumer AV, lighting and power supply manufacturers.

Holland:

Component and Material Consumption in Holland: FY 2017:

Electronic component and material consumption in Holland is value added and application specific and includes some unique products with emphasis upon line voltage devices, high frequency connectivity and high components and materials for medical test and scan equipment. Lighting ballasts are also an important part of the Holland high tech economy and of course, this is really describing the footprint for one of the key electronics firms in Europe-Royal Dutch Philips Electronics.

United Kingdom:

The United Kingdom has the massive BAE Systems at 16.3 BB British Pounds ($25 BB USD) and is Europe's largest electronics company and consumer of electronic components and materials. Rolls Royce Automotive Division and Aircraft Engine Division are also important customers for electronic components and materials and the bespoke nature of consumption for components; especially those operating at high temperatures >200 degrees C for defense aircraft should not be overlooked for future growth within the country. There is also the large Delphi automotive electronics plant located in the UK as a major consumer of electronic components and materials. GKN PLC is also a major consumer of components and materials for automotive and aerospace subassemblies.

Further investigation finds a large-scale operation for automotive production in the UK at 1.723 MM cars and trucks for 2016- a new record high. More cars are now being exported from Britain than ever before, the result of investments made over recent years in world-class production facilities, cutting-edge design and technology and one of Europe's most highly skilled and productive workforces. Ten brand new car models began production in the UK last year, nine of them from premium brands which has helped make the UK the second biggest producer of premium cars after Germany and the third biggest car producer in Europe.

Major Automotive Customers Driving Capacitor Demand in the UK:

Castle Bromwich/Halewood/Solihull Region:

This region experienced a 10.3% increase in car production in 2016 for the Land Rover factories producing 544,000 vehicles. All Rover vehicles are produced in this region.

Sunderland Region:

Massive Nissan factory produces 507,000 vehicles for 2016 including high electronic content electric and hybrid electric designs.

Oxford Region:

The MINI plant in Oxford produces 211,000 vehicles and is owned by BMW.

Burnaston-Derby Region:

Toyota has pulled back production on this plant in 2016 as the models seem unfavorable, but at 180,000 vehicles it is a large factory.

Swindon Region:

The Honda plant in Swindon produces some 136,000 vehicles.

Ellesmere Port Region:

Vauxhall, producers of the Astra in Ellesmere Point, grew 39% in output in 2016, to 118,000 cars produced. The Astra is a $20,000 USD compact car with 4 doors and an extremely powerful engine.

Additional Factory Locations:

Aston Martin (Gaydon); Bentley (Crewe); Caterham (Dartford);

Switzerland:

3% of European electronic component and material sales are in Switzerland and go to Asea Brown Boveri for industrial electronics and utility grade switchboards and switchgear apparatus as well as smart metering products. The strength of Switzerland's electronics industry lies in a small range of niche products, with high added value. Control and instrumentation, and consumer goods (mainly watches) are the two sectors of industry, which dominate Swiss electronics output. Watch production is the major industry accounting for 33.5% of total production with control and instrumentation having a 28% share in 2016. During the year a number of companies looked to utilize overseas operations to offset the ending of the peg for the Swiss franc against the euro.

Other important end markets and customers are as follows- **Bernafon** (hearing aids), **CS Customer Care & Solutions Holding AG** (EMS), **Endress+Hauser (Instruments), Enics (EMS), FM Acoustics (Audio), Logitech (Computer Accessories), Revox (Audio), Saitek (Game Accessories), and Studer, (Audio).**

Ireland:

The Irish market for electronic component and materials is also an important high-tech manufacturing base in Northern European.

Major Transplants and Home-Grown Electronics Companies in Ireland.

The Intel Factory in Leixlip, Kildare; is the largest manufacturing plant for Intel outside of the United States; Cisco is also in Ireland in Oranmore, Co. Galway. Dell is a major computer vendor located in Ireland (The company, still the third-biggest technology firm in Ireland by turnover, but it has largely switched to R&D); there is also EMC storage (bought by Dell), also operating in Ireland.

Key Irish Companies In The High-Tech Economy

Ireland is a "Friend To Technology" and has created a fragmented and diversified electronics Industry focused in power supplies, switchgear, industrial automation and some communications and display technology. Here is a list of key high-tech companies operating in Ireland-

- Anord Control Systems Ltd: Design and manufacture of low voltage switchgear and control solutions
- Associated Rewinds (IRL) Ltd: Remanufacture of traction motors for the locomotive industry
- Ceramicx Ireland Ltd: Manufacture of ceramic based infra red heaters and accessories
- Convertec Ltd: Design and manufacture of customised and standard power supplies
- Data Display Co Ltd: Manufacture of electronic information display solutions
- Datac Control International Ltd: Design and manufacture of remote monitoring systems,
- E & I Engineering Ireland Ltd: Design and manufacture of bespoke switchgear and electrical busbar products
- E M Con Systems Ltd: Emergency lighting, lighting equipment and systems manufacturer
- Eblana Photonics Ltd: Digital lasers for optical transmission applications

- Eurolec Instrumentation Ltd: Manufacture of electronic measuring equipment
- Excelsys Technologies Ltd: Supplier of high efficiency, low profile, user configurable modular power supplies
- John Byrne Conveyors: Design and manufacture of conveyor systems
- mSemicon Teoranta: Specialists in motion control, audio,
- medical device, communication, industrial and environmental technologies
- Novum (Overseas) Ltd: Manufacture of retail refrigeration equipment
- ONG Automation Ltd: Delivery of "wet process" automation
- Powerplex Tech: Innovative power supplies for the electronic security industry
- Sigtec Ltd: Design and production of wireless-based safety and communication technology
- Western Automation: Specialist suppliers of Residual Current Devices (RCD) technology to the electrical industry

Sweden:

The Sweden the high-tech economy is focused on large scale industrial automation equipment at related Asea Brown Boveri factories. This is also the home to the high-tech telecommunications switches made by Ericcson- the largest manufacturer in the country by far of high-tech related equipment.

Key Swedish Customers Buying Components and Materials

Elekta in Sweden is an important manufacturer of medical test and scan equipment and a key consumer of value-added and application specific components and materials in the region. Axis Communications is an important security company. FLIR Systems also has an operational factory in Sweden for the production of security equipment. There is also the Tank Radar Operations of Emerson.

The Southern European High Tech Economy:

The Southern European High-Tech Economy

The Southern European high-tech economy supports a market rich in automotive, telecommunications and space electronics.

France:

The French high-tech economy is one of the largest and most diversified in Europe and reflects an advanced technology base requiring similar components and materials.

Key Manufacturers in The French High-Tech Economy

France is a major producer in the fields of medical (Soren), Thales (Defense- satellites) defense (EADS), aerospace (EADS), energy (Alstom, Schneider LeGrand) Alstom Trains and telecom infrastructure (Alcatel). In Telecom infrastructure- France has many advantages, with a world leader, Alcatel-Lucent, and several major players, including integrators such as Thales, and companies operating on specific sectors (Gemalto, Oberthur and Technicolor). France can also count on innovative SMEs boasting advanced technologies, especially in the optical sphere

The French electronics industry continues to be dominated by the production of fixed and wireless communications infrastructure equipment. The country is also a leading producer of radar, navigation and defense electronics which accounted for 25% of the total output of electronics in 2017.

Importance of The Automotive, Industrial and Infrastructure Markets in France

The largest sector for electronics is automotive-France is home to two major automaking companies- Renault and PSA Group whose output is about 2.1 MM cars and lite trucks making France the fourth largest producer of cars in Europe. The support of the car safety and convenience electronics and hybrid electric projects in France would be the largest end use market segment in the region, followed by industrial electronics and power transmission and distribution equipment and telecommunication infrastructure equipment, each segment is where France maintains global brands,

Spain:

Automotive Industry Dominates Spain's High Tech Supply Chain:

The automotive industry in Spain remains a viable end-market for electronic components and materials in the country. Spain produced 2.7 million cars which made it the 8th largest automobile producing country in the world and the 2nd largest car manufacturer in Europe after Germany. In all, there are 13 factories located in Spain which are supported by a thriving local car components industry, including rapidly growing Spanish multinationals such as Gestamp Automocion (Steel) and Grupo Antolin (Car Interiors).

The major automotive factories in Spain are as follows-

- Daimler AG (Vitoria Plant)
- Ford's largest plant in Europe (Almussafes Plant)
- General Motors: (Figureulas)
- Nissan (Barcelona)
- PSA (Vigo)
- Volkswagen (Pamplona)
- SEAT (Mantorell)

SEAT is the sole active Spanish brand with a mass production potential and capability of developing its own models in-house. Today it operates as a subsidiary of the Spain is also the home to EPCOS' massive film capacitor production factory in Malaga.

Other Spanish Markets for Electronic Components and Materials:

The remaining consumption of capacitors in Spain is centered around satellite components and communications. Other major customers in Spain for capacitors are Fagor (Digital TV and EMS for auto and large Home Appliance), and TRYO Group for aerospace and defense electronics. The TRYO

Space Business Unit supplies advanced microwave equipment and subsystems from VHF to Ka band to the world leading satellite manufacturers and Space Agencies for a wide range of applications, such as Telecommunications, Navigation, Space Exploration or Earth Observation. There is also Telvent, which is owned by Schneider, and which manufacturers industrial automation equipment and line voltage equipment.

Automotive Electronics Factories in Spain:
Delphi has a large presence in Spain but it remains mechanical in nature and consumed very little in actual electronics, instead, producing switches, mechatronics and plastic injection molded parts for cars. The SAS Plants in Pamplona seem electronic component intensive and look like a key state customer for components and materials for cockpits and shifters.

Italy:
The Italian market for electronic component and materials boasts some 1,200 associated companies. The Italian Electronics Industry amounts to 63 billion Euros of turnover (of which 29 billion of Euros in export).

The Italian Automotive Industry (FIAT):
The largest market for electronic components and materials in Italy is the automotive industry which produced about 1.1 MM Cars and light trucks in FY 2017, and of that about 90% came from one customer- FIAT. FIAT brands include Lancia, Ferrari, Maserati and Alfa Romeo.

Other Industrial and Communications Markets for Electronic Components and Matreials in Italy

Ducati Energi is an Italian power industrial company that is based in Bologna Italy. This is a focal point of power and electrical manufacturing in Europe and this company also manufactures many advanced sub-assemblies captively.

Olivetti is a manufacturer of typewriters, computers, tablets, smartphones, printers and other such business products as calculators and fax machines. Today Olivetti is branching out in additive manufacturing. We consider Olivetti to be in the "Computer" end-use segment in Europe.

Radio Marconi is also one of the largest manufacturers of radio communication equipment in Europe and a major consumer of electronic components and materials.

Portugal:

Portugal assembles foreign cars and trucks with production of nearly 200,000 units annually including Volkswagen's "autoEuropa" and Groupe PSA Factories.

Key Automotive Customers for Capacitors in Portugal

Electronic component and material consumption is largely to the automotive space and the VAW factory and PSA factory respectively.

3.0 Structure of The High-Tech End Markets In Europe:

Automotive and Transportation Electronics:

The automotive electronics end-use market segment is the largest portion of the European regional market and accounted for 34% of European electronic material and component consumption in FY 2017. It is the largest market for electronic components and materials in Europe and impacts or defines the supply chains in multiple countries.

Automotive electronics include engine control units, ABS cards, SRS electronics, car stereos, HVAC systems, driver information and diagnostic systems, powertrain electronics, HEV integration, door locks, seat motors, interior and exterior lighting, instrument clusters and related electronics. The combination of an increase in electronic content per automobile, coupled with increased unit sales of cars and light trucks globally has increased the automotive sector in Europe.

The automotive markets for FY 2017 in Europe are active and robust and have been a considerable engine for growth. Demand for the 3.2 MM high end automobiles produced at BMW, Mercedes and Audi in FY 2017 turns into a large demand for electronics and capacitors as these markets are capacitor intensive.

Industrial Automation, Motors and Drive Assemblies:

The industrial electronics end-use market segment is a key portion of the European regional market and accounted for 18% of European high-tech component and materials consumption in FY 2017.

Motor Run and Drive Assemblies:

The motor market has traditionally grown in accordance with new home building globally, which traditionally drives sales of large home appliances, such as air conditioning equipment and refrigerators. The market for large home appliances and HVAC systems is driven primarily by new home sales; but aftermarket sales are also influenced by a quest for improved energy efficiency.

Paumanok estimates the global motor and motion market at $30 billion USD in 2017, with motors accounting for $18.0 billion in value, and drives accounting for $12 billion in value. In the motor markets- fractional horsepower motors accounted for 75% of global consumption value in FY 2017, and integral horsepower motors accounting for the remaining 25% of consumption value. The AC capacitor motor market, or those motors that consume AC motor run capacitors totaled an estimated 185 million units produced worldwide in 2017 with a market value of $4.0 billion USD. Europe accounts for 14% of the world's output of motors. The two major producers in Europe are ABB and Siemens, and to a lesser extent Compton Greaves.

segment will continue to underperform due to increased competition of Smartphones, which are displacing the needs for separate cameras, video devices and MP3 players and even game consoles. While we remain bullish on some products, such as Wii and Playstation® and Xbox for the future, other devices such as cameras, video and MP3 will continue to decline in the wake of Smartphone consumption worldwide.

This market segment declined steadily due to continued pressure by smart phones making certain consumer AV products redundant and unnecessary.

The top customers in the world for consumer audio and video imaging equipment include Samsung Electronics (18%), Panasonic/Sanyo (17%), Sony (14%), LG Electronics (8%), Philips (6%), Pioneer (4%), Hitachi (4%) and other (29%).

Large Home Appliances:

Large and small home appliances also make up a portion of consumption for components and materials in the European market with manufacturers such as Robert Bosch having key positions in white goods.

Medical Electronics:

The Medical Markets in Holland, France and Germany:

Medical components and materials comprise 6% of the European high-tech economy and are largely situated in Holland, Germany and France.

Medical test and scan equipment and medical monitoring pumps and compressor electronics are a large market for high voltage, high frequency and high reliability components and materials at Philips in Holland. The French market for medical devices is also interesting, with the country playing a major international role in the market for medical implants through Sorep. Medical markets also comprise about 2% of the large German market at Siemens AG.

Telecommunications Infrastructure:

Key Telecom Markets Driving Specialty Components and Materials in Europe: FY 2017:

While the European market has shed itself of almost all handset production over the past 15 years, the telecommunications infrastructure markets in Germany and France and Italy remain active at Siemens, Alcatel and Sasso Marconi as well as at Thales in France.

Products included in the telecom sector are landline and cordless phones, GPS devices, telecommunications infrastructure equipment such as switches, routers and repeaters; PBX equipment, wireless base stations, cable modems and various forms of commercial satellites and related communication equipment.

Computers

Computer Assembly Markets in Ireland, Poland and Italy:

The computer segment in Europe is largely different than what we see in China, but there are assembly houses in Ireland and Poland (Monitors) as well as in other regions where the computers are industrial in nature such as at Olivetti in Italy end-use market segment is a key portion of the European regional market and accounted for 6% of European materials and components consumption in F 2017.

Lighting Ballasts and Power Supplies:

The Ballast and power supply end-use market segment is accounted for 5% of European components and materials consumption in FY 2017.

The producers of lighting "components" differs somewhat from the major lighting "assembly" manufacturers. The point of sale for the electronic component manufacturer is the "ballast" or "driver" producer (a ballast in turn is very similar in design to a power supply). The major manufacturers of lighting ballasts and drivers includes Philips (Advance Transformer, EBT and AXA subsidiaries combined); Panasonic (which owns Vossloh-Schwabe); Siemens AG (which owns Osram), Tridonic ATCO, ADLT, Toshiba, Hella, Hubbel and Fulham Ballasts).

Defense and Aerospace Electronics:

The defense electronics end-use market segment is for electronic components and materials is very active in FY 2017. Major customers include Thales and EADS in France, and BAE in UK.

Power Transmission & Distribution:

The Power T&D end-use market segment accounted for 4% of European component and material consumption in FY 2017 and producer Asea Brown Boveri in Switzerland and Sweden is the major global brand in the space, competing against General Electric and Eaton Power globally.

ABOUT THE AUTHOR:

Mr. Dennis M. Zogbi is president and CEO of Paumanok Publications, Inc., a market research company located in Cary, North Carolina. specializing in market research studies, consulting, mergers and acquisitions, conferences and seminars with emphasis upon Paumanok Publications, Inc., a market research company located in Cary, North Carolina specializing in market research studies, consulting, mergers and acquisitions, conferences and seminars with emphasis upon passive electronic capacitors. Paumanok Publications, Inc. has a 30 year reputation and has 300 customers worldwide in the field of market research on Capacitors, Capacitors, inductors, circuit protection and electronic materials. Paumanok Research engages in off-the-shelf market research reports, single client research related to new product development, due diligence for mergers and acquisitions and for establishing business growth for passive capacitor companies worldwide. Paumanok Publications, Inc. also owns Passive Capacitor Industry Magazine with global circulation of 14,000. Mr. Zogbi engages in single client research related to new product development, due diligence for mergers and acquisitions and for establishing business growth for passive capacitor companies worldwide.

www.ingramcontent.com/pod-product-compliance
Lightning Source LLC
Chambersburg PA
CBHW071731170526
45165CB00005B/2239